LEAN CULTURE:

How to simply implement Lean thinking in your organization

INTRODUCTION

1984

Known most notably for George Orwell's prediction of "Big Brother" government and the tyranny it would cast on all our lives.

While that didn't happen (not totally anyway), another reason, literary even, for taking note of 1984 was a book called

"The Machine that Changed the World" by Dr. James Womack

Dr. James Womack of MIT, had spent the previous 5 years and $5M researching the automobile industry and trying to find which manufacturers were in the best position for future growth and who might be the leader going forward in making the world's cars.

What he found instead, was a way to organize any complex processes in the most efficient way yet achieved.

He found this at Toyota and he called this system—Lean.

From that initial start the careem of Lean would go into not only all manufacturing operations in any industry, but transactional office work processes, medical field, aviation-anywhere two or more steps are involved in bringing a purchased value to a customer.

Lean can be applied damn near anywhere.

To get a full appreciation of Lean and where it is today in modern society it is worth a short trek through history to provide a little context.

HISTORY

Dr. James Womack was educated in political science. He has a B.S. in that field from the University of Chicago and his Ph.D. in political science as well from MIT.

He was not an engineer and not generally known as such as opposed to many MIT graduates as it being at the apex of technology, science or mathematical complexity worldwide. Nevertheless his impact was felt among those disciplines probably as much as any other initiative over the past one hundred years.

His dissertation was done on a comparison in the industrial practices in Germany, Japan, and the U.S in 1982. (The importance of this will be made clear shortly). As part of that he spent five years and $5M traveling the world and looking at the manufacturing plants and the operations in hundreds of plants in those countries (and many others as well).

Womack say that Toyota, Japan's largest automaker, had a manufacturing process that was different from the U.S. auto makers he saw in Detroit. They still had assembly lines and used basically the same type of parts. But how the operations of assembly moved from one station to another was different. Even how the workers interacted with their bosses and with one another, even as far as how they actually thought about their jobs, their company and the problems they encountered were different.

In short the entire culture was different. Not just different due to an Asian-American, East-West type difference; although that indeed was part of it. But an actual approach to situations and problems was remarkably different.

The Toyota method was much more efficient. They produced more units with less effort and a higher margin while having fewer defects than the U. S. and even their German counter parts.

The word he used to describe this superior efficiency he called "Lean". Just that-Lean. Simple, accurate and to the point.

Lean as in "lean and mean". Meaning less fat, less bloated. Streamlined.

Notice Lean is not an acronym. I have found that in some plants usually union shops, many of the workers have come to think that Lean is an acronym. As in:

L-less

E-employees

A -are

N- Needed

The thinking being that if Lean is all about improving efficiency then if a process takes ten employees to complete it and it becomes more efficient to where now eight can do it— therefore two people lose their jobs.

What we will see later is that in Lean thinking just the opposite is the case.

Dr. Womack went on to not only teach but evangelize the world with the ideas that Toyota had and how they did things. He wrote several more books on the subject such as

"The Toyota Way"

"Lean Thinking"

"Learning to See"

"Lean Solutions"

And he founded the Lean Enterprise Institute in Cambridge Massachusetts

https://www.lean.org/WhoWeAre/

A nonprofit organization that educates and trains the world on Lean Principles.

At this point it would be logical for one to ask, "How it is that Toyota was able to become so efficient?

In fact during Dr. Womack's dissertation, the three particular countries (Japan, Germany and the U.S.) are a good juxtaposition on the history of Lean.

All three countries were involved in World War II. However Germany and Japan were on the losing end, and had their manufacturing base and infrastructure practically destroyed. Post- war they were literally starting from scratch in an effort to rebuild and rejoin the world economy.

The U.S. on the other hand was not only a victor but the war was not fought on American soil (absent Pearl Harbor attack). Their manufacturing base was not only intact but had expanded bigger than ever to support the war effort.

Interestingly enough, this difference is what later led to Lean being created and developed in Toyota while taking years to gain hold in the U.S. and other countries and companies.

The reason being, is that when you are down, at your lowest-rock-bottom state, you have nowhere to go but up and have to act somehow, in ANY manner to make things better.

The U.S. was in a diametrically opposite position. They were the main manufacturers of the world. Even the other Allies such as England, France, etc. had the war fought in their back yard, so they too were behind the U.S. in manufacturing capability.

If you wanted a new car, refrigerator, washing machines, TV (as they became available) or any product you almost assuredly had to buy it from the U.S. since they were the only makers "in town" so to speak.

America had practically a monopoly on manufactured goods worldwide, and with the massive rebuilding effort going on around the globe, the U. S. base companies got rich and grew even larger.

And what usually happens when you are at the top with little competition and little reason to change?

You become bloated and develop some bad practices that are covered up by gross revenue and profit expansion.

America became a literal fat cat.

Japan however HAD to change. Had to find a way to quickly get back in the game and grow their economy. They went looking for ways to improve and improve rapidly.

Ironically it was an American that helped Toyota achieves their renaissance.

Enter Dr. Edward Deming.

Edward Deming was born in the good ole U.S. of A in 1900. He was educated in electrical engineering at Wyoming University as an undergraduate and got a PhD from Yale in physics and mathematics.

He started his career working for the U.S. Census Bureau. He helped developed the sampling technique used by the Bureau of Labor Statistics in the late 1920's. While there he wrote books championing the Statistical Process Control (SPC) work of Dr. Walter Shewhart of Bell Labs who specialized in applied statistics in industry especially Design of Experiments (DOE).

 It was while working for Gen. MacArthur in Japan on the Japanese census after WWII that he was asked by Japanese businessmen to speak to a group on not only SPC but his whole theory of management that he had developed and written about over the years. It was over the course of 4 years that his theories and views were heard and accepted by Kiirchiro Toyoda the president of Toyota Motor Company.

From that the rest, they say, is history. Literally.

Toyota with little to lose and near rock bottom, implemented Deming's ideas and from that what is now today called the Toyota Production System and what Dr. Womack would eventually call Lean was born.

It seems that the ideas have an American origin in conception but brought to fruition by Japanese Eastern culture and companies.

Which brings us to today.

Enough of history. What is is about Lean itself that makes is so powerful?

PRINCIPLES OF LEAN

Although Lean can produce massive cultural and bottom line changes to a corporation or plant, the actual principles upon which it is based are actually very simple and easy to comprehend.

In fact Lean is based on only five general principles identified by Dr. Womack.

However, being easy to understand and being easy to IMPLEMENT-can be two entirely different things.

Think of it like an effort to lose weight. Most people understand quite easily that to lose weight one must eat less calories and follow a good diet while perhaps at the same time exercise or just move more to burn offf more calories.

Less calories in and more calories out.

Pretty simple.

Yet so many people fail at diets over time even if they are able to make some short term improvements. Losing weight, gaining it back, losing weight and gaining it back is the curse of far too many people in the U.S.

Though following a clean diet is based on simple concept and principles; the key is in the implementation.

Implementation is where it gets hard.

And so it is with Lean.

As previously mentioned Lean is based on only five principles:

   1. Define Value
   2. Value Stream Map (VSM) your process

3. Flow
4. Pull
5. Continuous Improvement (C.I.)

Only five.

These are what Dr. Womack was able to summarize when he spent so much time observing and examining what Toyota did when they went about their business.

Sometimes you may see these principles wording a little differently, maybe more wordy or lengthy, but at the end of the day put simply they are what you see above.

Simple enough in fact to be reduce to merely five literal words:

1. Value
2. VSM (ok not a single word but a phrase)
3. Flow
4. Pull
5. C.I. (two words)

There are few among us who can legitimately claim to not be able to commit these five words to memory and ultimately to understand what they mean.

In later chapters we will go through each of these principles individually in much greater detail. On a very high level, let's just describe slightly more detail in that:

- VALUE—anything the customer is willing to pay for. What do you do that makes the customer buy from you rather than your competitors. It's maybe less cost, more reliable product, more trustworthy on deliver, etc.

- Value Stream Map (VSM)—draw a picture of the process that you go through to deliver the value above to your customer. Identify what you do that adds value and what part of your steps that do not add value.

- Flow- try to get to single piece continuous flow. One operator per one part one at a time. This is opposed to batching which most companies do today. Now we can't always get to the ideal state but it is our goal.

- Pull- as opposed to pushing. One process step doesn't make or do anything till the downstream process step asks or "pulls" it from you.

- Continuous Improvement (CI)- always seek perfection. Have a culture such that every person from CEO on down to all hourly workers are always thinking of ways to improve their processes.

As mentioned before, we will go into a much more involved discussion on each of these principles.

In the interim I want to make the point that

-Lean is easy to understand

-Based on only five principles

-As a result can be taught to anyone and SHOULD be taught to everyone in a process, or plant, or company.

-Some people will be better at implementation of these principles than others- as with all things I suppose.

-But every single worker should be involved and all acting in concert working together.

What you will find as you advance your way through this book that one of the main themes of Lean is CULTURAL CHANGE.

If one just took the five principles and some of the various tools we will eventually go into like: takt time analysis, gemba walks, visual management, 5S program, and on and on- and we tried to apply or utilize these tools individually out of context what we would find that our processes really won't improve like we hope and we will never reach our potential and get the results from a true Lean implementation as we should.

Cultural change needs to happen from the CEO all the way down to every single hourly worker throughout the whole organization.

Now that may seem like a very large and daunting task initially but what you'll find that in as little as six months you can start to see some measurable substaintial changes. And after a year you may not even recongnize your old company. And after two years you will wonder how you ever lived with the old system you once were so use too.

So to really get the "bang for your buck" your goal should be to change your culture.

If you think about it, what gives you and edge over your competition?

They can still your ideas, they can steal your designs, they can even steal your products and people. But no other company, no competition can steal your culture.

This is why while many companies embark on a Lean implementation very few actually see big results like a Toyota does.

Other companies go into Toyota or pay outside consultants who have been there and maybe even once worked there, and based on what they saw they try to copy what Toyota does.

But if you just copy the "motions" you are mimicking shadows.

Capturing the style but not the substance.

The substance, the "meat" if you will, is all in the culture that Toyota has created over near sixty years by this time, is really at the core of why they make big improvements on a routine bases and other companies, especially western, U.S. companies will make modest gains-if they are lucky to have any at all.

So as you read your way through this book keep that in the forefront of your mind. That not only do you need to understand and then be able to use the tools; you also have to change your culture; i.e.—the way your management/leaders interact and work with your line and hourly people. The way your organization views "failure" and how to move past it. How open any person is to trying something new and communicating those findings to anyone else in the business regardless of their position or role.

While this may all seem abstract and academic at this point, it will be made more clear as we do a deeper dive into each principle and tool and go over real life examples and case studies.

Against that backdrop lets start delving into the core of each fo the principles.

1. **PRINCIPLE ONE: VALUE**

What is value?

Who defines it?

Value, like quality is defined by—the customer.

The customer and only the customer determines what is value.

The customer is usually referred to as the "end" customer or the person who will be the final user of your product.

But, there are also "internal" customers. What that means is that in any process, the next step down stream from you within a factory or company is your customer.

For example:

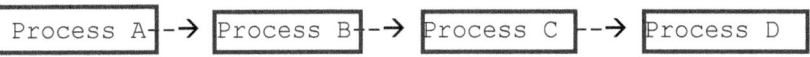

Process D is the customer of Process C, who in turn is the customer to Process B and so on.

Even if all these processes are within the same factory, same office building, same work cell, it is always the case that who ever is downstream of you is your customer AS WELL AS—the final end user who is paying your company for your product

That tells us who gets to establish value but it doesn't tell us what value is?

Taking the case of the end user customer since all of the processes ultimately work for the end customer, value is defined as:

"What the customer is willing to pay for".

Pretty simple.

To get more specific for a company maybe a better working definition is:

"Any action you do that changes the form or function of a product or service—that the customer is willing to pay for"

When I got to McDonalds, I, as the customer only want to pay for the hamburger. I don't want to pay for Ronald McDonald's costume, I don't want to pay for the CEO's second house in Miami, I don't want to pay for the Super Bowl commercials, I don't want to pay for the training of new employees, and on and on.

I only want to pay for the item I am going to use, in this case the hamburger.

All other activies McDonalds does is not what I want to pay for so in essence they are NON-VALUE added work.

All work of a company is either

-value added: changing the form or function such as what a customer will pay for

-Non-value added: all other work.

So cooking the beef, adding the oninions and cheese, etc. all add to the taste of the hamburger so I am willing to pay for that.

Everything else is non-value added.

Now just because something is non-value added does NOT mean it is UNNESSARY.

For example, Lean and Six Sigma training is non-value added and that is something I have spent over 20 years as a career doing. Even though it is non-value added I would hate to think it was also unnecessary!!

All work then falls into one of three types

1. VALUE ADDED
2. Non-Value added—but necessary
3. Non-Value added—but Unnecessary

A few examples

Non-value added but necessary:

-paying taxes(someone has to compute and make sure this happens)

-payroll

-advertizing

-marketing

-training

-environmental regulation compliance

None of the above are value added but likewise none of the above can be fully gotten rid of hence; they are necessary.
(Thankfully for me Lean and Six Sigma training is in this group)

Examples of non-value added work that is UNnecessary

-scrap (making bad product)

-rework (fixing bad product before you send it to a customer)

-inventory

-work in progress (WIP- a type of internal inventory)

Ideally in a Lean environment we want to eliminate if at all possible the non-value added unnecessary work and reduce the non-value added necessary work.

Even reduce the time and effort for the value added work if we are able.

But the primary focus should be on elimination of non-value added unnecessary work.

This type of work is waste and in the Lean world waste is categorized into one of eight types of categories

(1.)Transporation—moving material and product around

(2.)Inventory—build up of raw,final product or partial product (WIP)

(3.)Intellect—not getting enough thinking out of employees

(4.)Motion- motion of workers,i.e.—getting up to get equipment,etc.

(5.)Waiting-anytime a worker or down stream step has to wait on upstream

(6.)Over production- making too much of something

(7.)Over processing -to much inspection, processing, etc.

(8.)Defects- making bad products

This forms the acronym

T.I.I.M.W.O.O.D.

TIIMWOOD.

Often times you may see acronyms like

DOWNTIME. Which is Defects, Overproduction, Waiting, non-value added processing, Travel, Iventory, Motion, Employee intellect

Or

WORMPIIC. Waiting, Over production, Re-work, Motion, Processing excess, Inventory, Intellect, Conveyance

As you can see all contain the basic concepts even though the acronyms are different due to different wording and order.

In a Lean environment you try to reduce non-value added unnecessary work by eliminating or reducing the eight types of waste.

2. **PRINCIPLE TWO: VALUE STREAM MAP (VSM)**

Now that we know what value is, and what it is the customer particularly values about our services, i.e.-why they buy from us rather than our competitors; we need to pictorially represent how it is we go about providing that value.

Create a Value Stream Map (VSM).

A VSM is a paper exercise requiring no technology or even artistic ability. It is a simple visual representation put together by the very people (people, not one single person or a couple of guys. We will expand on this later) who are involved in carrying out the process on a day to day basis.

Usually this is done on brown butcher paper (the kind a butcher uses to wrap meat in when selling it to a customer at a deli) spread out over a wall at the site or area (if possible) where the process takes place.

Of course it doesn't have to be butcher paper any type of paper spread out in a long horizontal way will do.

Each operator or person in the process will put a yellow sticky (the 3M pads for example) on the paper representing the process they are involved in, then the next person and so on till you cover all the way from original input (customer order) to final output (customer getting their product.)

Below is an example of a VSM:

Value Stream Mapping Example:

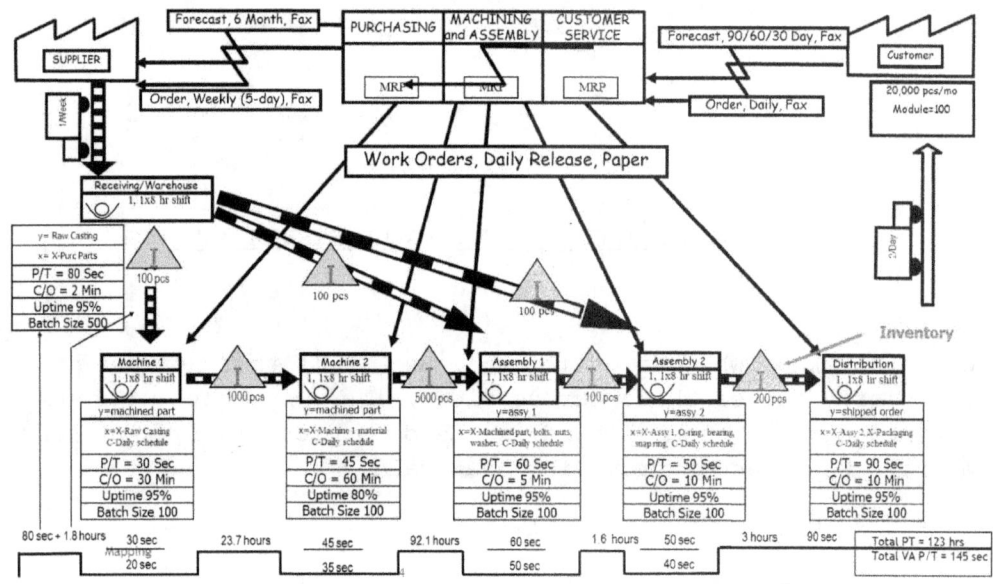

Notice that it not only captures each of the process steps, but the data involved in each process such as:

- P/T: process time or the cycle time which is the time it takes to do the step

- C/O: change over time or how long it takes to make changes to machinery to go from making one part to another part. For example if this were a paint machine and you were painting car doors red, and next you wanted to paint them blue, obviously you would have to clean out the machinery to get rid of any red paint residue before introducing the blue.

- Uptime: how often is the process running and how often it's down

- Batch size: how many items are you performing at a time

Etc.

At the very bottom in a "time ladder" that captures how long the entire process takes. In this case 123 hours.

Along with that it determines how much of that time is value added vs. non-value added. This is done by considering the definition of value and applying the seven types of waste.

As you can see there is only 145 sec of value added time. Basically the sum of the two machining process steps and the two assembly steps.

It is not uncommon that an entire process has such a low percentage of value added time to total time such as this.

(123 hrs.)*(3600 seconds/1hr) = 442, 800 total seconds

145 values added seconds/442, 800 seconds = .03% value added time!!

Most of the waste is in Inventory wait time or the parts and materials waiting between each step. The worse offender being the 5,000 pcs (WIP) between the second machining step and the first assembly step causing a wait time of 92.1 hrs.

Something must change and preferably changes here first if this process is to be improved.

Notice this is a systemic problem. It will require the entire process team involvement including the scheduler of this process and even sales team to come together at one time (a Kaizen event. More on that later) to resolve this.

The main takeaway from this, is that once a team can create a map like this the problem areas can become visually and apparent. There is a transparency to this that is very important to a process improvement effort.

This is why it is very important that a VSM be done at the area of the process, allows everyone in the process to participate and LEFT hanging on a wall in the area for all to see and everyone to be involved in the solution and acting in concert to make improvements.

Once a current state VSM is complete, the team should want to visualize and create a future state VSM. That is a picture of how the process SHOULD or "ought" to look.

In the current example from the VSM picture above that may look something like this:

Visualizing the Future State

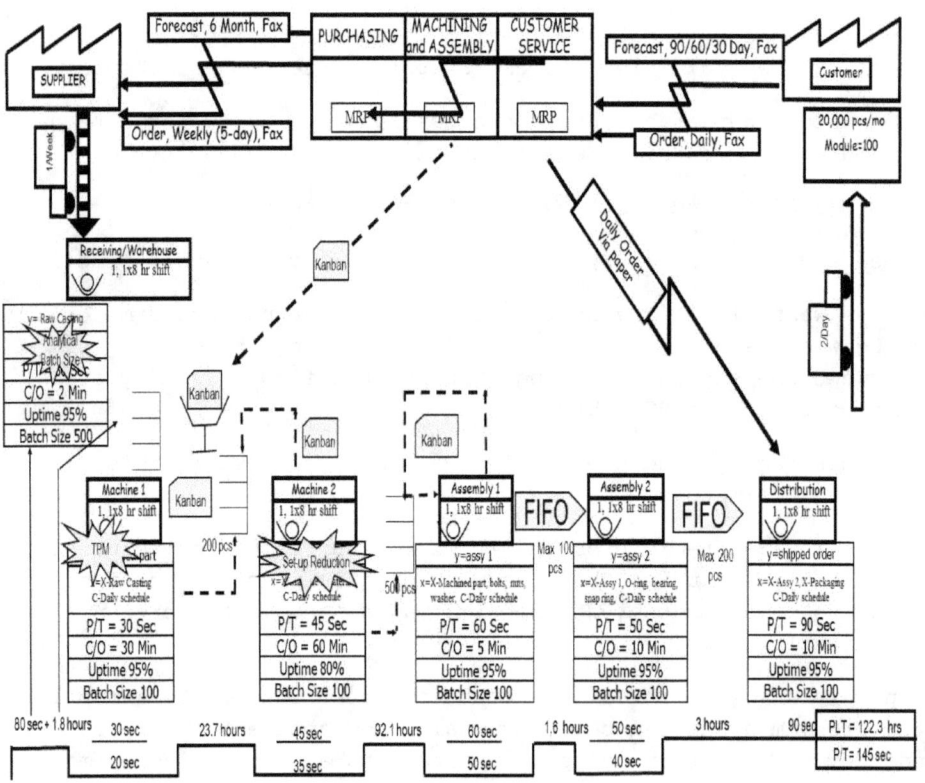

As you can see there are a lot of areas identified as places of improvements. We will discuss some of these later such as: FIFO (First in First Out) system, Kanban (to establish a Pull system), TPM (Total Preventive Maintenance), Batch size changes, etc.

The team should capture as many of these type of improvement ideas as possible.

The next step is to identify the type of work or projects that need to be done that will take you from the current state VSM to the future state VSM.

Projects like this usually fall into 3 categories:

1. Just Do it—that means literally what it says. A person or group of people should go directly from the VSM activity exercise and go to Gemba or the site and fix that which is identified.

2. Kaizen Event—we will discuss this more later but basically this is a group, team activity where operators, associates from the area in question, as well as "outside eyes" from upstream, downstream, marketing, sales, engineering, etc. come together for 3-5 days of concentrated work where they come up with a solution to the problem in question. The power of a Kaizen event comes in its focus. During those 3-5 days the people will walk away from their day-to-day jobs and just concentrate on this one particular problem. Too often problems linger in organizations because no one is personally responsible solely for that problem. Usually due to the fact that a lot of problems have cross-functional aspects that no one person has complete control over unilaterally. But if you get those cross functional people all together at one time this issue can easily disappear. The Kaizen event is one of the most powerful problem solving method in all of the Lean fiefdom

3. Six Sigma projects—this are projects that require more in depth, deeper dive into data and analysis and will not be solved quickly or any time soon. This would be a good way to identify who should be trained in Six Sigma as a Green Belt or Black Belt based on who works in the area in question. Too often organizations semi-randomly send people for what is often a long (3 weeks) and can be expensive (can range from $2-15K) training only to work on projects that may or may not be what the organization most needs working on. A VSM is a good tool to tie-in Six Sigma projects and training with the overall processes of a plant or site and will give more bang for your buck in the ROI of training and time spent on Green Belts and Black Belts.

Other projects like capital investments, major equipment movement and re-arranging, etc. should be considered LAST. All other improvement ideas should be exercised and carried out FIRST before any company commits a lot of time and money to try to add technology or more and better equipment to a process that has not already been optimized.

That would be one of the worse ways to invest money

From the VSM once you identified the various potential projects a good way to prioritize what to do first, is something called a "Benefits vs Efforts" (B&E) matrix.

This is where a Cartesian graph is created with "benefits" (B) as the y or vertical axis and "effort" (E) on the x or horizontal axis and the various project ideas are placed on this graph that gives the team an idea on what projects will give the most benefits with the least amount of effort.

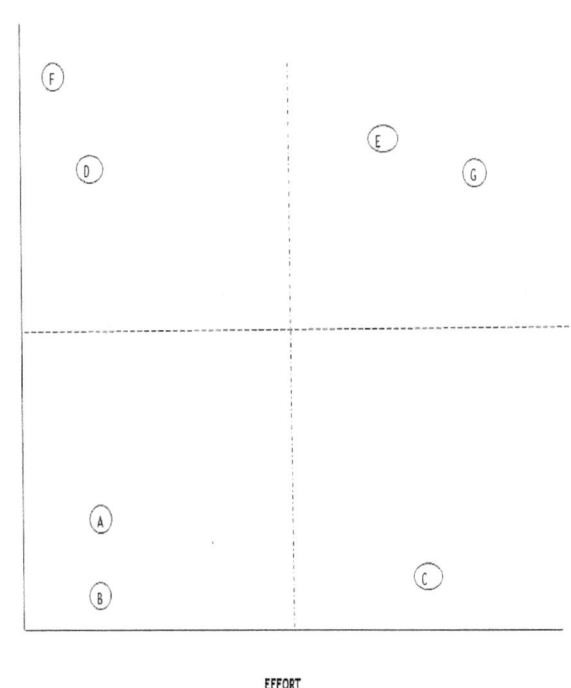

From the graph it should be obvious that projects labeled D and F since they give a lot of Benefits with little Effort. Likewise project C a team wouldn't want to do as it has the opposite characteristics.

The team would have to make a decision on projects A and B as well as E and G.

What should happen now is that a prioritized list would be:

F

D

A

B

Etc.

AS WELL AS identifying who (the team members) would be working on said projects also identifying key milestones such as start date, current status updates, etc.

This project list should be displayed openly at the very site, just like the VSM, where the process exists.

This transparency is very important in building a Lean culture as it keeps everyone informed of what the process is, the direction and vision of the improvement efforts, what particular projects are being worked, who is working on them, where they currently stand, what are the next steps, etc. etc.

Now one can start to see how Lean can catch on as a team effort and change the culture of the entire organization.

Value Stream Mapping is the foundation of all Lean work in any organization.

3. **PRINCIPLE THREE: FLOW**

Flow in the Lean lexicon refers to the movement of material, product or information in any process. Envision flow like a river running smoothly. Just like in the river metaphor if items block the river like a dam, a fallen tree or anything, the water will over flow the banks and flood the surrounding area.

You can see the same thing in manufacturing plants with a buildup of parts in the process. This is called WIP or Work In Progress. It is a form of inventory and as we've already mentioned inventory is considered a type of waste in Lean thinking. Recall the mnemonic TIMWOOD. (The "I" is inventory).

In Lean flow usually implies "single piece continuous" flow. This is opposed to "batch" movement of parts or product.

What you see in a traditional manufacturing plant (this also applies to any office or transactional process as well) is that one station or operator works on several items at once. Such as completing their part of say 100 parts, or 500 and then "pushing" that downstream to the next station or person who in turn do their work on the same 100 or 500 parts and likewise push it downstream again to the next station and so on and so on.

Notice that this is "batched" in 100 or 500 groups at a time.

This runs counter to Lean thinking which likes things to move, as much as is possible depending on the product or service, one part at a time. That is the stations above should make one part, pass it to the next station, which works on it and then passes one part, and so on and so on.

Below is an example of a traditional work line in an average factory:

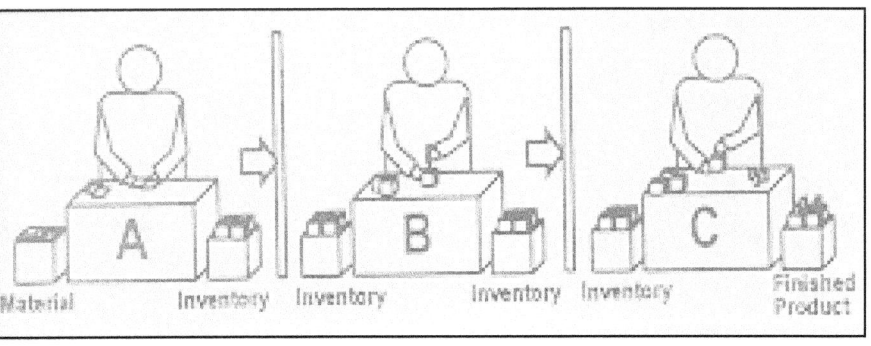

Notice the buildup of WIP between stations.

Compare and contrast that with a single piece flow set up:

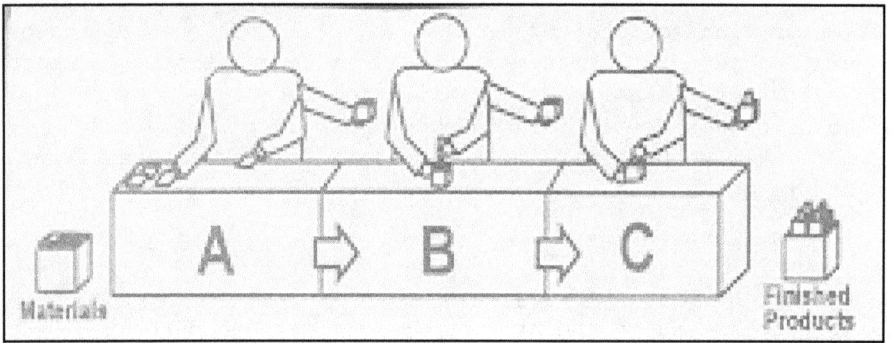

It may not seem like it at first but the later moves product along much faster.

At first glance to many this may seem counter intuitive. Many people think it seems less efficient to work on one part and pass it along rather than work on many parts like 100 or 500 and pass it on.

But in reality if you do the analysis you'll see that this is not the case.

In fact consider the scenario below:

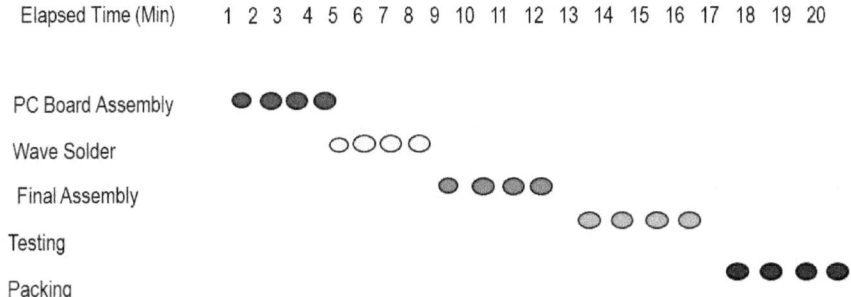

Lead times are long because each product in a batch must wait for the others to be completed before the next process can be started.

It takes 20 full seconds for the "order" of 4 is totally completed. This is while batching 4 at a time before passing it. I.E.—"PC Board Assembly" completes 4 items and then passes it to "Wave Solder" who then completes 4 items and then passes them in batch form to "Final Assembly" etc. all the way until "Packing."

Now what would happen if instead of batching in units of 4, we did single piece flow?

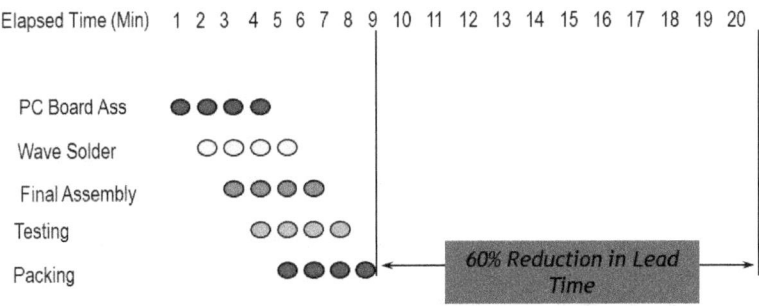

Now the order of 4 pieces is completed in 9 seconds. 60% less time.

How was this possible?

By getting rid of waste once again. In this case the waste is WAITING. Recall TIMWOOD. The "W" is waiting.

Before each downstream process was waiting, and in fact waiting a long time before any working material even showed up to their station. Downstream was waiting for the upstream processes due to the batch sizes that were needed to be completed.

With single piece flow this wait time is greatly reduced.

Also another benefit in single piece flow compared to batching is quality.

For instance if the first station were batching be it 4 units or 100, if that first station happened to have made a mistake and made all those units wrong, it would be 4, 100 or whatever bad parts before the next operation got a chance to notice that there was a problem.

With single piece flow the next operation notices a bad part right away. Therefore less waste in terms of defects (TIMWOOD, the "D") is encountered.

It should be clear now why the word flow is so appropriate as batching slows and interferes with clean smooth flow while single piece continuous movement integrates right into a great flow environment.

A good short working definition of flow that is commonly used is:

UNINTERRUPTED VALUE ADDED WORK.

Meaning that if operators in a process are working in an uninterrupted way and only doing value added work- that is with no waste- then that would be the epitome of good flow.

Every organization or process should strive to reach that ideal goal. To have all workers be it on a factory floor or in an office environment, working in an uninterrupted way and only doing value added work.

One logical question that many would have at this time would be: "if you want workers to work uninterrupted all day long, how fast should that do that, since no one can go full out fast as they can all day".

This is true. This is why Usain Bolt, the world's fastest man who holds the world's record in the 100m sprint; doesn't also hold the world's record in the marathon. He, and no one, can keep up an all-out pace for that long.

In fact the faster one usually works the less time they can do it and usually vice-versa.

What's the result of each station going at different speeds?

One big negative factor is the build-up of WIP or "Work In Progress" which is basically a type of inventory built up between stations.

One Lean tool that tells allows us to quantify this negative impact is

Little's Law: Lead Time = WIP/Exit rate

We will talk more about Lead Time shortly but for now just suffice it to say that it is the time between when a customer gives you their order and you give them their product.

The Exit rate is how fast a product comes off of your line

Take a look at this example:

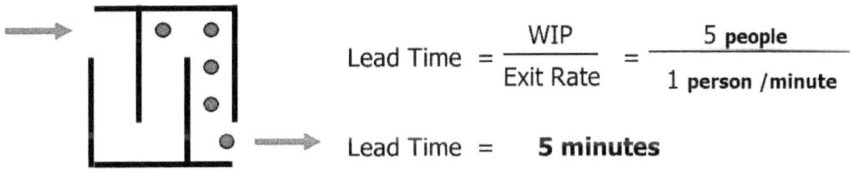

Think about the lines at Disneyland in March...

$$\text{Lead Time} = \frac{\text{WIP}}{\text{Exit Rate}} = \frac{5 \text{ people}}{1 \text{ person /minute}}$$

$$\text{Lead Time} = \mathbf{5 \text{ minutes}}$$

...and then think about them in July...

$$\text{Lead Time} = \frac{\text{WIP}}{\text{Exit Rate}} = \frac{13 \text{ people}}{1 \text{ person /minute}}$$

$$\text{Lead Time} = \mathbf{13 \text{ minutes}}$$

The Exit rate is how fast people can get on a ride so while that remains unchanged (unless they change something technologically with the ride itself- a big investment) the wait time or Lead Time is totally effect by the amount of WIP; i.e.- people in line.

The same is true in a manufacturing environment.

Take a look at the following scenarios:

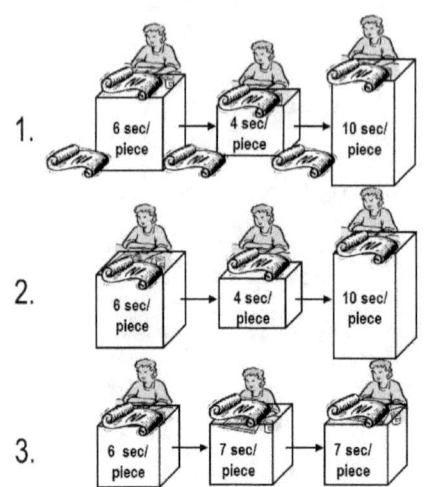

Present State
WIP = 6 units (papers)
Time Trap = 10 seconds
Capacity = 6 units/minute (1 unit every 10 secs)
PCT = 6 units / 6 units per min = 1 minute

WIP Reduction
WIP = 3 units
Time Trap = 10 seconds
Capacity = 6 units/min (no improvement)
PCT = 3 units / 6 units per min = 30 seconds
(50% decrease)

Capacity Increase
WIP = 3 units
Time Trap = 7 seconds
Capacity = 8.5 units/min (1 unit every 7 secs)
PCT = 3 units / 8.5 units per min = 21 sec

15

Just by reducing the WIP from scenario 1 to 2 you get an improvement in PCT (Process Cycle Time, in this case it's the same as Lead Time.) from 1 minute to 30 second.

By just re-balancing the work load between the last two stations (one is 10 sec/piece, the other 4 sec/piece. Divided equally they are both roughly 7 sec/piece) with the WIP now gone, you get further improvement from 30 sec down to 21 sec.

Notice that in none of the cases there is no technology advancement, no real improvement in the value added work needed, no addition of people or machines, etc.

All this is cost free improvement.

Also take note that WIP builds up due to one station moving faster than it's downstream customer.

This leads us to a logical question one must ask:

So how fast is fast enough then?

How fast should workers work?

This brings us to the concept of: takt time.

The word "takt" is not an acronym but a German word that describes "keeping the beat or pace", or "tempo". Think along the lines of "tick-tock, tick-tock".

It is used in German for the wand the conductor holds when directing an orchestra. And this is an appropriate description as a conductor keeps the woodwinds, the percussion instruments, the bass, etc. from all chiming in at any time they want to. They are all on pace at a certain time otherwise it would be a mess of noise like a bad heavy metal band rather than a well-received audible sensation like a classical music piece.

And so it is with a manufacturing line or any process involving many steps.

Takt time is the "pace of sales". Or the rate we need to produce in order to meet sales, which is basically customer demand.

At Toyota's largest North American plant in Georgetown, KY (first built in 1987 and located about 15 min from where I am writing this), a Toyota car comes off the line every 53 seconds.

Now does that mean it takes only 53 seconds to make a car? Of course not. It actually takes weeks to make a fully car.

But it does mean that somewhere in the world a dealership or someone is selling a Toyota car every 53 seconds on average. So to full fill that now demand to replace that Toyota car that has been sold- every 53 seconds a car has to come off that line.

The equation for takt time is:

$$\text{Takt Time} = \frac{\text{Available Time}}{\text{Customer Demand}}$$

The "Available Time" is the time a plant has to work in making the product.

If a plant works around the clock, 24 hours, 7 days a week then the numerator is maxed out at 168 hours.

Then takt time is only a function of customer demand and nothing else.

Notice that one cannot walk onto a factory floor and measure takt time by watching the operators work. Many people confuse takt time with "cycle time" which is how long a given operation takes for a worker to carry out that action.

Takt time only cares about how fast the customer wants the product. Nothing more.

Keep in mind that in the "Available time" that, if a plant does not work 24/7, and the line comes down for things like: breaks, lunch, maintenance, etc. then that time has to be taken into consideration in the "available time".

Consider this example:

A plant has a customer that wants 534 units of whatever, every day.

This plant runs one shift that works 8 hours. Every day the workers take a 5 min break in the morning, and a 30 min break for lunch.

What is the takt time?

This is basically how fast does each worker have to work in order to get the customer what they want?

Take the equation above:

Available time is found by—

8 hrs. /day = 8hrs * (60 min/hr.) *(60 sec/min) = 28,800 seconds each day

Subtract out the 30 min for lunch + 5 min break = 35 min = 2100 sec of non-work activity.

We are left with

Available time = 28,800 sec/day – 2100 sec/day = 26,700 sec

Putting that into the takt time equation gives:

$$\text{Takt Time (TT)} = \frac{\text{Available Time (AT)}}{\text{Customer Order Qty}}$$

$$\text{TT} = \frac{26{,}700 \text{ sec/day}}{534 \text{ units/day}}$$

$$= 50 \text{ sec / unit}$$

Which means that each operator must complete their operation every 50 sec in order to meet customer demand.

And what if they don't?

Then in this scenario they would have to work overtime, which is a higher labor cost; or have a second shift, etc.

So when someone asks "how fast should workers work?" (As we did above), then the answer is

"Workers should work at takt time"

Or at the pace of customer demand, the pace of sales.

No fast nor no slower.

Now would be a good time to define some of the times used when discussing a process and how they relate to one another.

As we've just defined takt time is the pace of sales or tempo of customer demand.

We've also stated that "cycle time" is the speed at which an operator can do a particular step.

What then should be the relationship between cycle time and takt time?

Ideally that should be

Cycle time = takt time.

If this true then we can meet customer demand without incurring any over time or working with a 2nd or 3rd shift or on weekends, etc. OR missing out on customer delivery.

Customer on time delivery is an important Lean metric expressed as

OTD= On Time Delivery.

A standard for OTD in most industrial settings is at least 90%. Anything less than that is considered bad and world class is in the 99% range.

We've also mentioned "Lead time". What is Lead time in relation to takt time and cycle time?

Lead time is the time it takes from when the customer orders the product until the time they receive the product.

It should be clear that:

Lead Time > cycle time

Since cycle time is the speed of any one, given operation.

In an ideal state where there is no waste the following should be true

Lead time = cycle time1 + cycle time 2 + cycle time 3 + …..+ Cycle time N

Where there are "N" operations being done.

In real life, as we will see, this too often is not the case.

Of course this is all true for when a line is already up and running and does not include the start up. What I mean is that the line is "primed" or already has work being done and in front of each station already. There is always a start-up time cost.

What happens if these times are out of sync?

Consider the diagram below:

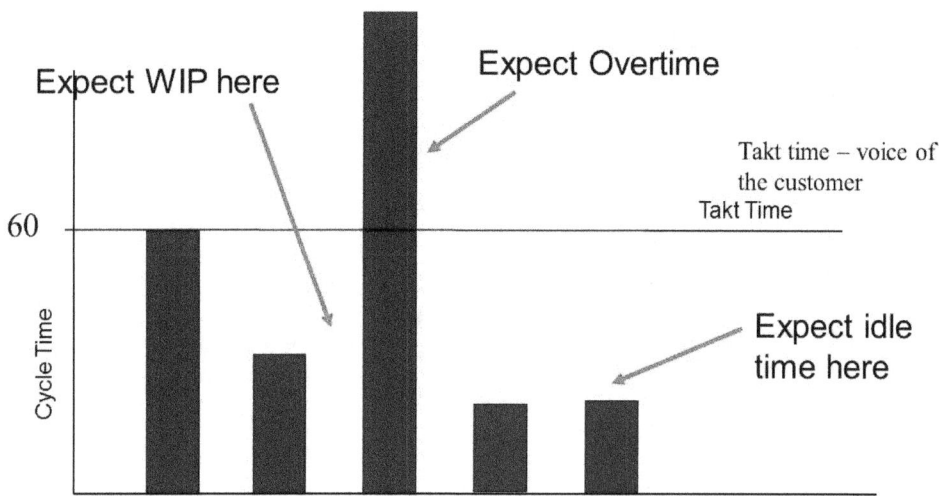

In this scenario 60 secs is the takt time.

The third station, station C, takes way longer so as a result of that, and that alone, there must be overtime in order to meet customer demand or our OTD will suffer.

In fact in any process with a series of steps the entire process can only go as fast as the slowest operation. Sort of like a chain is only as strong as its weakest link.

Also above there is a lot of idle time built in, in stations B, D and E. Those operators have a lot of time available to them.

It should make sense by now that this process should re-balance its work load (if it can. This is not always possible depending on the nature of the work of each station.) to have operators at stations B, D, and E perhaps help operator at station C.

A better approach would be:

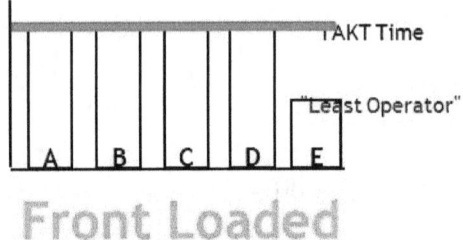

Front Loaded

Now stations A-D all work at takt time and this eliminates wastes.

Of course station E now has a lot of "free" time so what's to be done with him/her?

As we will discuss later, this person will now be a Lean leader and help coordinate and run Kaizen events, do lean training, work as a "water spider" (picture a rover who can move around to help others), etc.

The reason you want the work "Front Loaded" as is labeled here, is that the last position needs to have the idle time cause it will cause a "pull" (the next Lean Value we will discuss next chapter) that bring along and paces the entire chain of processing.

There are, however, other ways to re-balance workloads some with disadvantages and some with advantages.

Follow the various changes in the set up below:

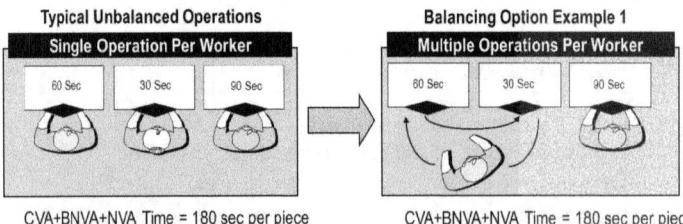

Process Balancing

Need improved productivity, do not need additional throughput

Typical Unbalanced Operations

Single Operation Per Worker

Balancing Option Example 2

Combine Operations b/t Workers

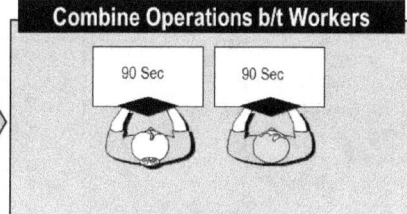

CVA+BNVA+NVA Time = 180 sec per piece
Bottleneck = 90 sec per piece
Output Rate = 40 pcs/hr
or 13.3 pcs/hr/worker

CVA+BNVA+NVA Time = 180 sec per piece
Bottleneck = 90 sec per piece
Output Rate = 40 pcs/hr
or 20 pcs/hr/worker

BUT NOW
The operators do not have to move, removing ergonomics and safety issues

Need improved productivity, do not need additional throughput

Balancing Option Example 3

Single Operation Per Worker

Multi Operation Per Worker

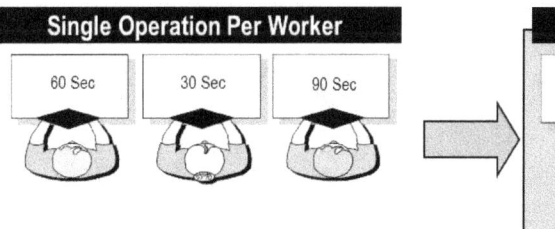

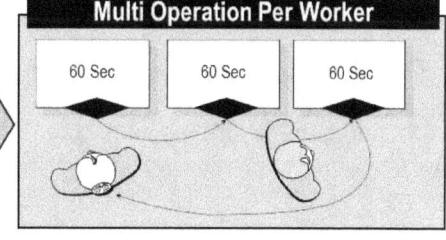

CVA+BNVA+NVA Time = 180 sec per piece
Bottleneck = 90 sec per piece
Output Rate = 40 pcs/hr
or 13.3 pcs/hr/worker

CVA+BNVA+NVA Time = 180 sec per piece
Bottleneck = 60 sec per piece
Output Rate = 60 pcs/hr
or 30 pcs/hr/worker

BUT again
requires operators to move a lot during the day - safety and ergonomics issue

Need improved productivity, and need additional throughput

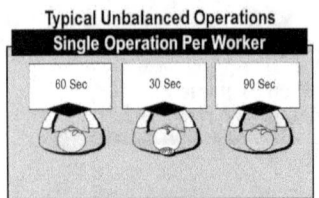

Typical Unbalanced Operations
Single Operation Per Worker

CVA+BNVA+NVA Time = 180 sec per piece
Bottleneck = 90 sec per piece
Output Rate = 40 pcs/hr
or 13.3 pcs/hr/worker

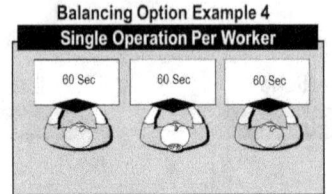

Balancing Option Example 4
Single Operation Per Worker

CVA+BNVA+NVA Time = 180 sec per piece
Bottleneck = 60 sec per piece
Output Rate = 60 pcs/hr
or 20 pcs/hr/worker

Operators don't have to move as much

Depending on whether ergonomics, man-power limitations, or efficiency are your goals each solution brings some different goal.

This entire chapter was about instituting as close as possible single-piece continuous flow as opposed to batching.

But in some cases batching inevitable.

For instance what if your plant is in Los Angeles and is supplied by a supplier in China where you get parts from a ship that arrived in a U.S. port after being on the water for four weeks? Obviously you can't put just one part on a ship for that long and hope to run single piece continuous flow from your supplier throughout your plant

Or if you have a chemical plant that has a reactor (Rx) that holds 1500 gallons of material or you have a furnace that runs best when it is completely full- in both those cases the size of the Rx and furnace dictate the batch by which parts and product need to be run.

In such cases single piece continuous flow doesn't work and batching is unavoidable. It only makes good sense to run to the batch size of the limiting process step.

Nevertheless the rest of the process steps should strive toward single piece continuous flow so as to always feed the physical constraint. You don't want to create an artificial constraint somewhere else and thereby have a Rx, a furnace, a dock or what have you waiting on parts. That would run counter to good flow.

The important lesson to take away is always feed a constraint and never leave it starved for parts, information, supplies or what have you.

Single piece continuous flow everywhere else is one way to make this happen.

Another way is PULL.

Which brings us to Lean value number four

4. **PRINCIPLE FOUR: PULL**

Pull as opposed to Push is the fourth principle of Lean.

A traditional manufacturing environment, one station in a process typical works at a pace as fast as it can or whatever is convenient and then just literally "pushes" the WIP or material downstream to the next station in the process regardless of whether that station is ready or not.

When you have a Push system you usually see the following:

- Outputs are generated as fast as possible.

- Expediting is the norm.

- Stacks of obsolete inventory are common

- Long list of projects "initiated", few completed.

- Despite piles of inventory, what the customer wants is seldom available.

Picture a snow plow pushing snow off a parking low

The snow piles up in big piles-although it is off the parking lot.

Just like the snow, in a manufacturing site, the work, or WIP (Work In Process) piles up. It piles up between stations and this (as we saw in the last chapter on Flow) slows down the entire process, increases the Lean Time, and causes the customer to wait longer for their order to be full filled.

But unlike the snow in a parking lot, WIP doesn't melt. It stays there and globs up the whole process.

Push, in Lean Thinking, is something to be avoided.

What then is a Pull system?

A Material Pull System means that downstream processes fetch from upstream processes only the goods that are needed, only when they are needed, and only in the required amounts.

Think "Just in Time".

A few characteristic about what a Pull system is and is not:

-NOT MRP (Material Replenishment Program like SAP or Oracle) based on Forecast

-NOT MRP with firm orders on MPS (make-to-order), because WIP is not limited

 Does not require an extensive computer system to keep work order inventory status correct

-not operations working blindly from one another

A traditional Push system then:

-Forecast Usage

-Large Lots

-Hidden Problems

-Waste

-Poor Communication

-Approximation

While a Lean Pull system is:

-Actual Usage

-Small Lots

-Visible Management

-Minimum Waste

-Good Communication

-Precision

A Push system does not limit or even care about WIP. Picture:

Notice the build-up in front of the secondary station on the right.

A Pull system deliberately limits and controls WIP in contrast:

Now take note of the clean, uncluttered WIP and how the work can move more easily at a steady pace between the two stations.

The governing principle of any Pull system then is:

Starts = Exits

The number of units entering into a system should equal the number of units exiting the system.

Or we only work on a new unit in a process when we have completed an older unit.

Hopefully that should be intuitive on some level.

The pictorial process shown below is your basic supermarket, ala Wal-Mart like.

Whenever a customer takes a product off the shelf puts into her shopping cart and when she checks out each product is bar scanned and this is how the bill is added up.

At the same time the bar scanning accumulates into a data base and at the end of the day, a signal is sent to the back area which shows what all products were taken off the shelf for that day.

Someone then takes product from the back; bar scans them, and refills the shelves in the store.

Likewise that bar scan from the back accumulates all the products refilled and it sends a signal to a distribution center which replenishes the back of the store.

The distribution center (as you maybe have guessed by now) also sends a signal to the suppliers in the supply chain about what to refill.

And the whole process is complete if done correctly.

You can follow all the numbered steps below that shows the entire fulfillment process.

Kanban A Simple Common Idea !
The Super Market

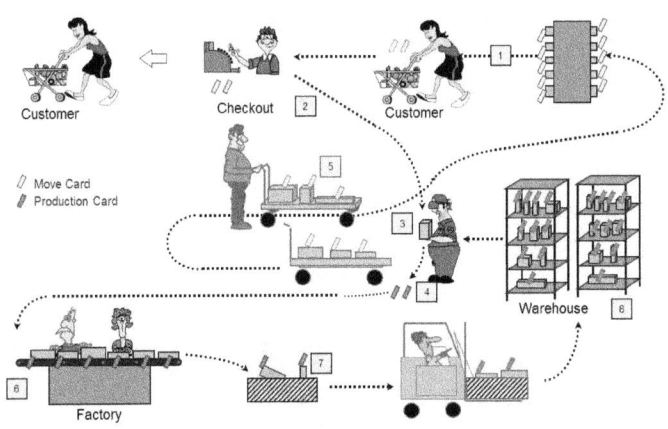

A much more simpler model would be your basic office vending machine:

Principle 4: Pull

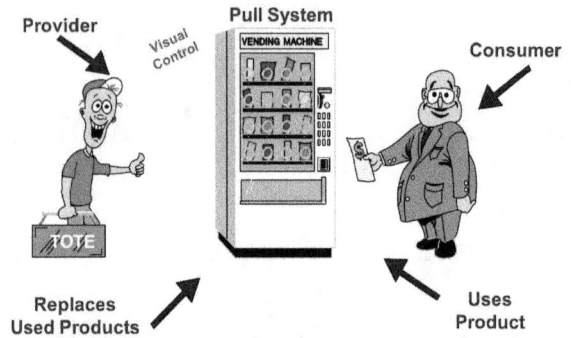

A typical vendor refills a machine once a week. They only refill the machine with only the products that had been taken over the course of the week. Nothing more and nothing less.

In Pull only what has been used is replenished and there's no need, at least in theory, for any forecasting, guess work, etc.

Pull goes hand-in-hand with making good Flow, value 3 in the Lean world, but a natural question might be at this point, how does a process make a Pull system work in an area that had always been Push before?

One key way is through a communication method called Kanban.

The word Kanban literally means "card" and it refers to using a card system to communicate to stations up and downstream in a process about what materials are needed.

When material is taken from one station to move downstream, a card is then used to pass UP stream to the previous station which signals that the material has been taken so that now the upstream step knows now to send new material to the original downstream station.

That is a very wordy explanation for something that is actually very simple as seen below

Kanban Squares

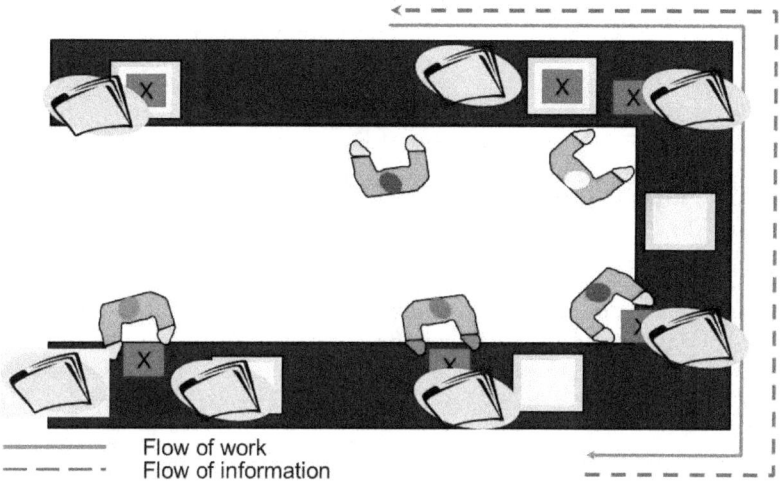

————————	Flow of work
— — —·	Flow of information

As material goes one way (clockwise in above diagram) the Kanban signal (information) goes the opposite way (counter-clockwise above).

This was created by Taichii Ohno at Toyota back in the 1940/50's and they used literal cards. Nowadays the Kanban signal is almost always electronic or digital in some way.

Some key characteristics of a good Kanban system is:

-a simple and visual system.

-good tool used for signaling between the customer and the supplier.

-transfers ownership back to the workers.

-Improvements come one part or one supplier at a time.

In order to make Kanban work well there are usually 6 general rules that need to be adhered too:

-Do not send defective products to the next process. (this is vital. If you follow this you can always get closer to root cause if there ever is a problem. This is lost if the problem is passed down the line)

-The consuming process comes to withdraw only what is needed.

-Produce to replenish only what is withdrawn by the next process.

-Parts must not be produced or conveyed when there is no need.

-Kanban cards must be attached to the actual parts or container.(this is still done today even in the era of digital electronics since floor operators don't often have a computer readily on hand and need to be able to "follow" parts on a factory floor)

-The actual quantity of parts in the container must match the number on the Kanban card.

There is an implication in a Kanban pull system that all parts will arrive on time when expected BUT what if that doesn't happen? What if a machine goes down, an operator makes a mistake, or as Toyota found in 1996; an ice storm hits and your suppliers can't deliver parts to your dock or trucks can't show up to take away your products via shipment.

What then?

It would be wise to build up a "buffer" or safety stock at strategic locations in the process. Obviously this needs to be kept as low as possible to make it work and this level needs to be calculated based on past incidents, since what we are describing is a type of inventory and as we've already learned, in Lean Thinking inventory is considered a waste or "bad".

Here's an example based on whether the customer's expectation is at or less than the Process Lead Time (PLT) of particular process:

Manufacturing Pull System:
Other Strategic Buffer Options

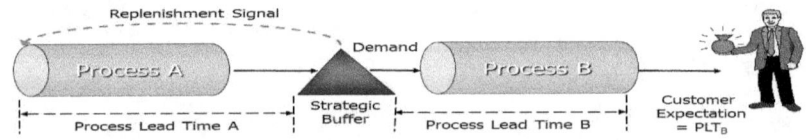

Customer expectation is the PLT of Process B only

Customer expectation is shorter than the PLT of Process B

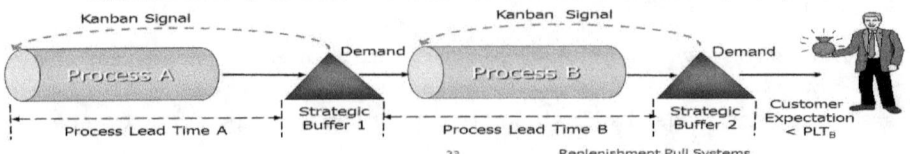

Replenishment Pull Systems

The strategic buffer allows the line to continue operating and keep good flow even if Process A or B has a "hiccup" or problem.

But be careful. Too large a strategic buffer and you basically just introduced WIP or inventory into your system that can slow it down and interrupt flow the very thing we are trying to avoid.

Make sure there are some key historical data and mathematical calculations around how big that buffer should be.

This is a good place to point out a key integration or marriage of Six Sigma tools to bring to bear in a Lean system.

5. **PRINCIPLE FIVE**: CONTINUOUS IMPROVEMENT

Continuous improvement or always "Seek Perfection" is our Fifth principle of Lean.

Toyota like all companies good at Lean know that they will never be perfect. Certainly not after only a few changes. With that in mind they have to be realistic and know that changes, constant change and changes are needed in even the most advance Lean facility.

Zero defects are the goal for Toyota but very few Toyota processes get there.

However that doesn't mean they don't stop striving for that.

One major way facilities can improve is by holding Kaizen events.

A Kaizen event is a concentrated anywhere from 1-5 day (usually 5) event where a group of people involved in a process (and a few upstream and downstream of said process, as well as perhaps sales, marketing and scheduling) gather together in one place at one time and work on trying to solve a SPECIFIC problem with the resources they have available.

The power of the Kaizen event comes from two things basically:

1. Focus—as the people/experts are honing in on one particular problem without distraction. The closer the group can keep to the integrity of this method the better will be the result

2. Experimentation—during the Kaizen event it is expected that the group will propose a solution and it will be TRIED in the plant and the result evaluated and if it doesn't fix the problem, adjustments are made and the process is repeated. Basically the PDCA (Plan Do Check Act—we will go into this in greater detail later.) applied in a few day period on one particular problem.

3. The fact that a group of people usually are able to produce or create a more optimal solution to a root cause problem than any one single individual.

Ideally a plant should do a Kaizen event once a month or bi-weekly depending on the size of the facility and the number of people working there. This cadence is important as over time the site will realize there is a formal, rigorous problem solving method that will happen on a periodic basis that everyone will eventually participate in to help solve the various problems of the site.

Think of the monthly Kaizen event, in the same way as you do preventive maintenance. Taking down a machine to do maintenance even when it is not broke, is a way to increase its life. Do preventive maintenance prevents or at least delays a catastrophic breakdown of a machine in the near future.

Also it gives maintenance people and even operators more experience and knowledge of how the machine operates.

Kaizen events are like that as well. Kaizen events are in essence a group problem solving effort. If you do this on a periodic basis even when a major problem does not present itself, it can stave off the bigger problem that one day may occur. Or prevent it all together.

Likewise it gives the people and the organization more experience at problem solving and in the long run this maybe the most important feature of a monthly Kaizen event.

This is one of the most important factors in building a Lean culture as almost any tool in the entire Lean kingdom.

The word itself, Kaizen, is obviously of Japanese origin. In Asiatic characters it is:

Change for good.

A few common areas Kaizen events usually concentrate on are:

-Reduce cycle time

-Reduce waste of wait time

-Reduce number of process steps

-Reduce inventory

Everyone should eventually be involved in a Kaizen event. With that in mind it is important perhaps to know what a Kaizen event is NOT:

-A tool to be used to reduce headcount!

-A way to do pet projects.

-Just an expert activity.

-A win - lose situation.

By all means don't use a Kaizen event to find a way to reduce the number of full time employees (FTE).

But, if you have a Kaizen event on a process let's say, that involves 10 people. Now say that you make it more efficient so now only 7 people are needed to run the process.

What then happens to the other 3?

The correct Lean answer is—those people are re-deployed. One area they could be is the new Lean Kaizen coordinators in working to organize future Kaizen events.

The reason you DON'T want to just fire those 3 people is that were you to do that, then how much participation would you get in the next Kaizen event, once the rest of the plant sees their coworkers lose their jobs?

It would be the worst kind of irony to ask 10 people to get together and think of a way to improve a process and then turn around and let a few of them go!!

Kaizen events require the input and thinking brains of the various people in the process. Why then with that in mind would you then run off those vary brains that you need?

Lean is all about creating "finder/fixers" or "improver/growers" or all around problem solvers. Any process that you improve and free up capacity then use it to grow the business. That should be the goal of any business anyway ultimately.

If you learn nothing else from this book learn that. Do not lay off or fire people due to process improvements from Kaizen events.

This illustrates some of the differences between the way a Lean company like Toyota looks at things and the way a traditional company does.

In a Lean company they:

-Look for problems. If there are no problems, there are no improvements. If you don't see a problem you probably haven't looked hard enough or really pushed the process hard enough to the point of breaking

-Problems are opportunities

-People are not the problem

-Teach people to be good problem solvers

Lean companies legitimately look at problems as ways to improve their businesses. In fact they are very suspicious of processes that claim to have no problems.

A good phrase to drive this home is a mantra used by University of Kentucky basketball coach John Calipari. Coach Cal is known for having what is called "one and done" players in college which are college players that play for one year at the University of Kentucky (UK) and then are good enough that they leave to be drafted into the NBA the very next year.

As a result of this Coach Cal has new players every year.

Since he has little time to turn these new players into a cohesive team he needs to quickly figure out their potential and talents and in order to do that he needs to see them-FAIL.

That may seem counter intuitive at first gleam but there is a logic to it.

By pushing the players to the point that they fail he knows exactly where their boundary conditions are. In fact that is the ONLY way you can truly know that.

Based on that criteria he wants them to try many things, and to do it quickly-fail, improve then fail, improve, etc. etc.

His mantra then is:

-Fail early

-Fail often

-manage your failures.

This is a theme in Lean companies as well. They foster an environment where workers are asked to figure things out, and to do that they need to change something and when they do that it is invariable that something will FAIL. But not only is that is ok it is expected and desired because it is from that failure that the future improved state is reached.

Consider the great inventor Thomas Edison. He attempted over 1,000 ways to invent the light bulb and got them wrong over and over again.

Then finally he got it right.

But you only need to get it right once in order to have a light bulb from now on.

The same is true for the polio vaccine, the atomic bomb; Apollo missions almost all human advancements. They all are built on past failures.

It behooves companies then to foster a "safe" environment. I don't mean safe as in free from physical injury-although that is vital- but I mean safe as in free from excessive criticism or negative consequences when a worker fails at something when he is trying to improve that very something.

Now keep in mind the last bullet item of "managing your failures". This means conduct your experiment in such a way that when you fail, it is not a catastrophic breakdown.

When you teach your child to first walk you don't do it at the top of the stairs.

Later when you teach them to ride a bike they are GONNA fall for certain. So you don't teach them on broken glass do you?

When you teach them to drive you do it in an empty parking lot or a little travelled road.

This should make sense right?

Bottom line don't be afraid to fail, don't let your coworkers and employees think that they can't fail, and don't avoid or shy away from failing.

Weightlifter and powerlifters (I competed in powerlifting for 10 years and once had a 470 lb. raw bench press) know that in order to keep lifting more and more weight they got to find and eventually go past their personal record (PR) in a particular lift.

When they fail on that lift, based on where in the movement they fail they can figure out what their weak point might be. E.g.—if my bench press fails at the bottom I know that my chest is weak. If it fails at the top then it's my triceps, etc.

Again notice the absolute essential nature of having failure as part of that success story.

To build a Lean culture, a business has to change from:

-Results without regard to process (always be process oriented)

-Functional

-Who did it?

To an environment where:

-Process and results

-Total system

-What happened? (as opposed to "who did it?")

From the old way of:

- Things viewed in a vacuum
- Manage events, things
- Separate people
- Individual performance valued at expense of system
- Bursts of energy to get things done n-o-w
- Competition

To the new way of:

- Things viewed in context
- Value harmony, stability
- Manage people, relationships
- Bring people together
- System performance is paramount
- Steady and smooth workflow
- Cooperation

	Old Paradigm	New Paradigm
Measurements	End result only	Trends of improvement
Support Staff	Critical of shop floor	Serves shop floor
Problems	Rejected/hidden	Treasures/communicated
Solution Focus	People	Systems/processes
Information	Restricted/closed	Shared/open
Methods	Static/routine	Changing/improving
Management Approach	Crisis	Preventive
Supervision	Inspector	Coach
Employee Development	Do as told	Learn by doing
View of People	Costs	Assets

6. <u>ADDITIONAL LEAN TOOLS</u>

If you get nothing out of this book but the understanding and ability to implement the 5 Lean Principles, that is a big win. But there are other tools in Lean that can be brought to bear to improve the efficiency of a company's processes

This other tools do not contradict and indeed supplement the Lean Principles. In many ways they make them easier to implement and carry out on a day to day basis.

Let's take a look at some of these tools.

<u>VISUAL MANAGEMENT</u>

Anyone who's ever watched a football or basketball game or any sporting event on television, can attest that it is possible to come in late to your house, turn on the TV and immediately see who is winning, what's the score, how much time is left, and a variety of other important variables needed to determine why the game is being played as it is at that very moment.

In a sporting event the Visual Management is basically the Scoreboard.

Well in a factory or company there should be a "scoreboard" as well.

Every single person on a factory floor, in a office process, or anywhere else in a company should be able to answer the question:

"Are we winning or losing?"

In business what does this mean?

It means:

-are we making product at the takt time and will be able to deliver the correct amount of good product to the customer

-are we on track to make our gross sales for this quarter to meet our year end goal

-are we on plan to make the quality improvement plans we set for ourselves this month

Every company, every plant, every process has goals. Goals it needs to be profitable, to make enough profit to meet Wall Street's expectations, goals to improve customer demand.

All companies have large goals that can be broken down into smaller goals derived for each of it's business units and ultimately factories and factory processes.

How then can every operator know if they are "winning or losing" against that goal for their operation without a "scoreboard" of some type?

The answer is they can't.

Visual Management answers this by proving production boards. These boards usually are hour by hour tracking of how many units have been made compared to how many should have been made to meet the goal.

It should also answer the question of if we are behind pace, why are we behind and what is the problem solving being done to bring us back on pace?

And like turning on a game in the middle and being able to determine what is going on, a good Visual Management Production Board is the scoreboard that anyone from the CEO, plant manager to all the operators can walk up to and see right away

"are we winning or losing"

Visual Management boards can also be in the form of monthly and quarterly metrics. In fact no company should be surprised by a quarterly result. If they have good Visual Management then should be able to track metrics monthly, then weekly, then daily and even hourly so they know well in advance what their quarterly (and by extension yearly) results will be.

This should also be communicated to Wall Street investors, owners and others who have a fiduciary interest in the company.

Visual Management must be updated regulary and TRANSPARENT. Team work, which Lean is really based on, is completely dependent on transparency.

Many of you have no doubt seen monthly and quarterly boards in the entrance of many companies but all they amounted to was life-less pieces of paper on particle board. Few people look at them and no one acts on them.

This is a very impotent Visual Management system.

Rather than just a voluminous report a trully productive Visual Management system should be like a stop light. At a stop light you don't just notice the light color. You ACT on it. If it's "green" you go. If it's "red" you stop.

A Visual Management system that does not cause an operator or process to act a certain way is not a very effective Visual Management system.

It should be an intrical part of the decision making process of workers and not just "wall paper" suppossedly decorating a hall way in the entrance in a poor way to impress customers and clients.

5S

One way to make Visual Management systems work better is with a 5S program.

5S stands for 5 words all beginning with the letter "S".

1. Sort
2. Set
3. Shine
4. Standardize
5. Sustain

This words are in a specific order as well. The best way to think of this is to consider this scenario:

Say you want to clean out your garage so that you can work on your car. You don't have a lot of time so you want to be able to do this as quickly as possible.

The problem is your garage (like a lot of factory work stations) is a mess. Clutter and tools every where. Not to mention just the filth.

When you (or a worker in a factory) comes upon such a situation the first thing they want to do is

SORT—meaning sort out the tools you immediately need from those tools you don't. So wrenches, screw drivers keep nearby. Golf clubs, hack saw you can move some place else. If you have a doubt as to whether you should move something or not, a good Lean technique is to place a red tag with a date time on it. Then every time you use that tool or piece of equipment update the date and time. If a week or month goes by and the red tag hasn't been updated, then, well you don't readily need that tool and it can be moved to somewhere else

SET—now that you have only the tools you need, set them in a certain place and always put them there when not in use. Picture a shadow board showing the outline of that tool such that when the tool isn't in that place, it's lclear on the shadow board that it's gone. Also should be a sign in/sign out sheet so that if it's not in it's place on the board, you can look at the sheet and know who has it. Every tool that you have SORTED and identified as being readily needed now has a specific place for it to go. A place for everything and everything in it's place.

SHINE—Now that you have only the tools you need and they all have a specific place to go, it's time to clean the place up. Or make it Shine. Even with a place for everything and everything in it' place; a dirty work place is not very organized. Clean and clean often. In fact clean the work area after every time you do work there. In a factory that would mean after every shift. In an office environment that would mean at the end of every work day.

STANDARDIZE—Now that we have just the tools we need, and each tool has a correct place to go and we've cleaned the area, we now need to make a Standard way to do this every single time whenever the area is used by anyone else. Anyone who works in the area or even anyone who walks by the area should be able to see how the area should look. This makes the solution of how the area should look systemic and not dependent on any one person.

SUSTAIN—What we mean by sustain is that we need to have a way to insure that over time, even when all the people in the process have moved on to other jobs or other areas, that this particular work space will always be in this condition. This is usually done by having third party audits at regular intervals or by surprise. Pictures taken periodically to compare against the standard. Continual training of the workers and operators. This way the momentum doesn't end.

What is the purpose of a 5S program? To make for a clean work environment?

While that is desirable the real reason for 5S is for one thing:

Immediately be able to tell "normal" from "abnormal"

That's it.

Many people are under the mistaken view that 5S is just a "clean up" program.

Not so.

It literally is a PROBLEM SOLVING program.

The originator of 5S, Taichi Ono one of Toyotas best engineers designed to aid in problem solving. The idea being that the closer to the root cause that you can get to a problem the easier it is to solve. And the sooner you see the problem the closer you will be to the root cause and the easier it will be to solved.

In a 5S environment when something is abnormal it is readily apparently. When that's the case a team can get to the root cause more quickly and come up with a solution much faster.

Taichi Ono was famous for his "Ono Circle". He would take a new engineer and bring him to the middle of the factory floor, draw a circle on the floor and have the engineer stand it for an entire shift. At the end of the shift he would come back and ask the engineer to describe what he saw. If the engineer could not identify all the problems that Ono knew existed he would have him stand in the circle again the next day.

It was important to him and is important to Toyota and any Lean company to be able to identify problems quickly as they occur.

This is best done in an area that has implemented a 5S program.

It's not just cleaning. It's problem solving.

TOYOTA KATA

In 2009, Mike Rother, a renowned Lean expert who has done a lot of research independently and with the Lean Enterprise Institute, wrote a book titled "Toyota Kata". In this he describes an important aspect of the problem solving culture at Toyota that to many outside observers is easily missed due to its subtlety.

Many companies embark on a Lean journey but never really get the results they should or want and one of the reasons is that their culture didn't' change. The implemented various Lean tools thinking that that was all they needed.

What is missing is this concept of PDCA Kata that Toyota does that Rother describes in his book.

If you are familiar with martial arts you have probably heard the word "kata" before. A kata is a choreographed previously established well rehearse action. The purpose of practicing a kata is to have it automatized so that the action is automatic when someone attacks you. Since you have little time to think out a response during a surprise attack, by practicing it over and over, your response will be immediate and reflexive and thereby making your more likely to defend yourself successfully.

In this context, Toyota does this when they approach a problem.

They have an automatic method they go through every time they encounter a problem so that it is automatic in every worker.

This is very powerful in altering the culture of an organization in making it into a problem solving, Lean culture and as a result their processes are much improved.

The nuts and bolts of this problem solving kata in Toyota has three main prongs

(A.) _PDCA_—this stands for "Plan, Do, Check, Act". This is basically the scientific method brought to industry. Everyone one of us at some point whether it was in a high school biology class, first semester college chemistry 101 or what have you, have encountered the scientific method of problem solving. Doesn't mean you have to be a lab coat wearing scientist to employ this just your regular thinking rational human being. This was popularized by Edward Deming (previously mentioned in earlier chapter). Basically you

PLAN: _Identify the problem_—this maybe the most important step. Abraham Lincoln once said "if you give me an hour to cut down a tree, I will spend 55 min sharping the ax". Likewise the famous GM inventor Charles Kettering (inventor or the electric starter used in all cars today) also said "A problem well defined is a problem half solved". Putting your problem into a short pithy declarative quantifiable sentence or two tees it up well for finding a solution.

A problem statement should have certain qualities to be productive.

i.   Clear
ii.   Concise (specific)
iii. Measurable
iv.   Time stamped (when did it happen)
v.   A standard and deviation from the same
vi.   Should not have a solution stated or implied in it

Two or three shortly worded sentences at most. Do not have a long meandering paragraph as a problem statement.

Consider some examples of real life problems statements:

a. *Chronic supplier problems are causing OTD issues*

   While this is short it is not quantifiable, it has an implied solution (supplier problems) and no standard and no mention of how much we deviate from it

b. *Accident strains are too high due to lack of back supports used by dock workers*

   Again short but still not measurable and has an implied solution

C. *Market share for TV's is down 24% since Jan 2015 in North America*

   Notice how this is much better. It is measurable, time stamped

   And just tells us about the problem doesn't imply a solution.

DO: once we have a well-defined problem we need to come up with a plan on how we will attack this problem. Producing a hypothesis that would explain the root cause, prioritizing all the potential hypothesis that maybe at play; we need to have all this laid out formally before we take a single action. Then and only then will we try an action. This is the DO part of PDCA. We are actually going to do something.

Check: Once we do something or take an action, now we will check to see if it solved the problem or not. In real life the first thing we do probably will not solve the problem unless we are very lucky. This will no doubt take several iterations. If the problem isn't solved we are back into the PDCA cycle. Back to the DO stage. What this means is that we will try a different action against the hypothesis or try to attack a different hypothesis all together. We stay within the PDCA cycle until a solution is finally found. Look at the picture below

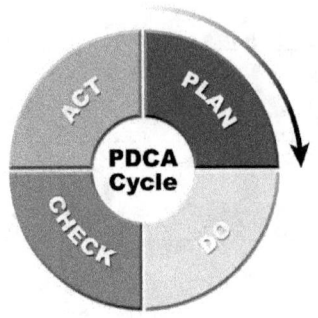

This represents the reiterative aspect of the PDCA cycle.

ACT–finally ACT. This is what we do once a solution is found. What do we put in place to make sure this problem never resurfaces? How do we make a systemic fix that will outlast the current people involved? A good solution is in the process and is not people dependent. Think of the 1980s San Francisco 49ers. They won Super Bowls with Joe Montana. When Montana left they brought in Steve Young and were able to win another Super Bowl. This was because their "West Coast Offense" was a system that worked well in the NFL at the time and wasn't' totally dependent on one man like Montana to make it work. Same thing with a problem in a company. When this is done communicate to the rest of the company so that everyone now knows the solution.

Overall then the PDCA looks like:

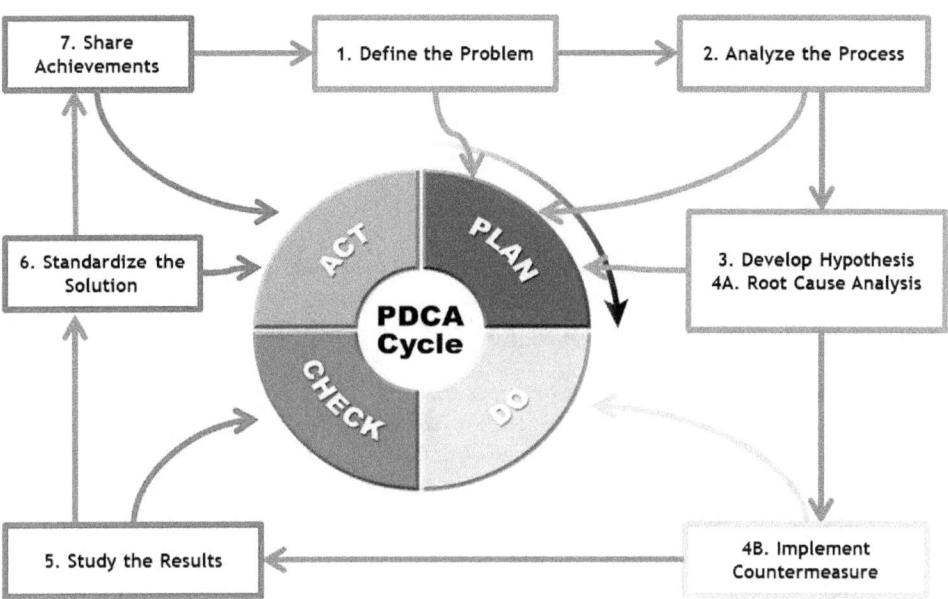

In the context of problem solving kata, it is the PDCA cycle that Toyota and other lean companies do continually when they encounter problems. Managers and supervisors mentor people who work for them in the use of this PDCA method. It's important to realize that managers don't solve their problems. They let them solve their own problems in their areas. But they just insure they are using a valid cogent reasoning method that is the PDCA cycle when they tackle the problem.

The chief way they do that is with the next tools we will discuss; the A3.

(B.) A3—PDCA is the problem solving method Toyota uses, an A3 is the tools that they use to formalize and communicate that method between one another and the rest of the company.

One of the most common questions I get on Lean tools is what is an A3? Where does that name come from?

Many are disappointed to learn that the answer is not some new software or application. It's just the size of a sheet of paper.

The story that gets told and it maybe apocryphal is that in 1984, when Toyota started their first joint venture with a US car manufacturer at a plant near Claremont, California; they had about 100 engineers on the ground and one group was haven't a persistent problem that they could not solve.

After many frustrating phone calls across the Pacific, the design engineers back in Japan told the engineers in the US, to just take the biggest sheet of paper they could fit through a fax machine and then fax over everything they currently knew about the problem.

The biggest sheet of paper that could fit through a fax as 11"x17" and is what we in England and the US call a "A3". From that the A3 method of presenting a problem was born.

While an A3 is just literally the size of a paper, it is formatted in a certain way so that a problem can be presented in an intelligible way to management or anyone else in a quick easy to comprehend manner.

Being just one sheet of paper it can be posted on a board or a wall in a work area or factory cell and right away any operator, the plant manager or even the CEO can come to the process vicinity and immediately see what the problem is, what is being done to fix it, the current status, etc.

An A3 is a visual representation of the PDCA problem solving cycle. The one sheet of paper is usually divided up into 8 (or 9) areas that correspond to different stage of the PDCA cycle.

Below is a picture of a commonly used A3:

PDCA	Team Name:	Topic:	Team Members:			Start Date			
0. Link to Strategy:			4A. Root Causes:	4B. Selected Countermeasures:		Who:	Start:	Target:	Finish:
1. Problem Statement:									
Understand the Circumstances									
Goal Statement:									
PLAN						**DO**			
2. Cause-Effect Analysis:			5. Results:						
3. Data Collection (Hypothesis Testing):			6. Sustain:			**CHECK**			
80%			7. Recognize / Share Achievement:			**ACT**			

Notice that 80% of one's time should be spent in the PLAN phase of PDCA.

One sheet of paper. 7-9 (varies by companies) segments or blocks it's broken down into. All in one way or another corresponding to a step or actions taken towards the steps in the PDCA cycle.

Using this a mentor can work with another associate and have the associate walk them through their thinking using the PDCA method and the mentor can give his thoughts or guidance on whether the associate is true to the logic of the PDCA.

This helps the associate with their problem solving skill as well as keeps the mentor and anyone else up to date on the status of the problem.

In many truly Lean companies you can find dozens of A3s displayed on walls and boards of all most any and all of their key process areas and cells.

This can be used in an office environment as well as a manufacturing floor.

Recently I have noticed many companies using A3's but not in a way that was intended. They use them as some type of reporting tools, for group presentations in meetings to executives. This is fine to do but it is not the real purpose of the A3. The A3 is most powerful when used on a local level at the place where the problem is occurring and used as a living problem solving document and tool.

If you or your company is using A3s and you have very long problem statements and several pages (a page or each section of the A3) then that is a pretty good sign that you've lost focus on why A3's exist in the first place. This is usually a sign that you need to take a step back and re-examine your Lean program.

When properly used an A3 is a fantastic tool for changing the overall culture of a company on their Lean journey.

(C.) *Individual Daily Kaizens*—the last part of the Toyota Kata, is something called Individual Daily Kaizens. We mentioned Kaizen events earlier which were group oriented (7-12 people usually) working 3-5 days exclusively on one particular problem. Usually we do a Kaizen event once a month.

If you'll recall the last principle of the 5 Principles of Lean is "C.I." or "Continuous Improvement".

While monthly Kaizen events are improvement opportunities, the fact that they are monthly tells us that this is continuous. As in literally continuous as in every hour of every day continuous.

Now there's no way we can do Kaizen events every hour of every day, so how do we reach this continuous state?

This is done by the inputs and thinking of the associates, operators and each person who works in an area of the company.

What is meant that every day and if possible several times a day any and all workers should be thinking of ways to improve their areas that they are the subject matter experts (SME).

This is captured by the Individual Daily Kaizen.

Each person should be able to create 1-3 ideas every day that could possibly improve their work areas. These ideas should be posted of displayed in a place; allowing for transparency; and leaders and supervisors should go over these lists every day.

In a Lean plant like Toyota in Georgetown, KY which has over 7500 (10,000 if you count temporary and contract workers) employees is capable of producing 50,000 ideas from their workers each year!

Now most of these ideas are not implemented. In fact most of the ideas will ultimately be proven to be bad ideas.

However, some aren't. And more importantly a good idea can come from a partial aspect of one bad idea, combined with a partial aspect of another bad idea, etc. Rarely does one person produce one fully developed good idea that can be immediately implemented. More than likely a good idea will come from combining 3 or more bad ideas in different ways in context they were not originally submitted for.

Nearly all human great ideas from Edison's light bulb (he had over 14 engineers working for him), Einstein's General Theory of Relativity (he used the Michealson-Morely experiment combined with the Lorentz transformations to come up with this) and thousands of other ground breaking ideas were usually the combination of several other ideas most of them at face value were bad ones.

That's how progress occurs.

And this is vital to the Lean transformation in a company.

GEMBA WALKS

Gemba means "where the work is" and a Gemba walk refers to a person, specifically plant/site leadership going to the location where the value added work is being done. That usually means the factory floor in a manufacturing plant. Or where the tellers are at a bank, where the nurses are in a hospital, etc.

A Gemba walk is when said leadership actually walks to and through the value added areas.

In a Lean culture plant managers, controllers, quality managers and everyone in a leadership position will walk the factory floor every day. Ideally twice a day. Usually once in the morning at the beginning of the work day and once in the afternoon at the end.

Most of a site's leadership meets every morning anyway, usually behind closed doors in a conference room in front of computers and overhead projectors and go over Power Point presentations or Excel spreadsheets.

In Lean thinking that meeting should be a mobile one where the team goes cell to cell or station to station on the plant floor and talks to the operators and supervisors at those value added locations about what is going on there.

What should happen is that an operator should explain to the group:

-what happened over night at the cell

-what the status is now and if they are "winning or losing" compared to the target on the hour by hour chart

-any problems or issues that exist in the cell

-go over any A3's regarding those problems

-go over any ideas or Individual Daily Kaizens that the operators in the cell has created or came up with over since the last Gemba Walk.

The Gemba Walk should take no longer than the previous conference room morning meeting that the team was already doing. It will take some time to get used to it, but eventually the goal is to add no more time than was taken before. But during this walk the team gets a direct look at what value added work is being done.

As some manufacturing sites are very large and it would be impossible for a leadership team to cover the entire plant in a reasonable amount of time. In a case like this then the plant should be divided up and the team would rotate through each cell during a week.

For instance they could cover assembly, welding, packing on Mondays. R&D, IT, maintenance areas on Tues and so on till the whole plant is covered over the course of a week.

Instead then of the team just sitting in at a desk in a CR going over Excel sheets based on data that they would have had to ultimately gotten from the manufacturing floor originally to being with. Now the team actually goes to that site and sees first hand where the value for the company is being added.

Plant and site leaders need to realize that even though they are "running" a plant, they are really there to SUPPORT the value added workers who are literally making the product that the customer is buying. This may come as a harsh reality to some in these positions, but in Lean thinking ego needs to be kept at the door.

When a plant is really doing well with this what it will look like is:

- Every cell will have a board that shows: Production (winning or losing), Quality, Safety, and Costs or PQSC and updated to the day

- A spot on the floor where each member of the team will stand

- A general knowledge of what the team is looking for and what to present to them

- A3 board someone can explain

- An idea board for Individual Daily Kaizens that the team can evaluate

So no more plant meetings behind doors in conference rooms.

In fact other than HR related issues and perhaps some company proprietary items, there's no reason that a Plant Manager and his direct reports shouldn't have their desk ON the plant floor rather than in a nice office at the front of the plant with a big oak desk and lots of windows and personal items.

Leadership is there to support the value added workers on the floor and should be as close to them as possible.

A Gemba Walk is one day to enhance that.

SUMMARY

To get the most benefit of a Lean program at your company, then you have to do more than just learn a few Lean concepts and apply a few tools from paper in class-room learning

Lean to be most effective needs to be a cultural change.

Cultural change takes a long time. To make it happen as quickly as one is able, you need to actively involve everyone in the application on a day to day basis of Lean principles.

Dr. James Womack numerated the 5 principles of Lean based on his research on Toyota manufacturing ideals. Those principles are:

1. Value---the customer defines value. They do so by what they are willing to pay for. So value is "anything you do that changes the form or function of a product that the customer is willing to pay for". All other work is non-value added. However not all non-value added work can be eliminated completely

2. Value Stream Mapping (VSM) --- you need to make a pictorial map of the steps you take to deliver the value mentioned above. This map should be produced by everyone involved in the process and displayed in an open and transparent way to everyone can see it at all times. The areas of waste should be readily apparent in a VSM.

3. Flow--- think of flow like a river running unimpeded. Flow can be defined as "uninterrupted value added work". The rate at which a process should flow should be equal to takt time. Takt time is the "pace of sales". How fast the customer is buying the product from you.

4. Pull--- this is opposed to "pushing" in that a process step takes no action and produces nothing until the downstream produce or the internal customer orders something. This keeps WIP or Work In Process down and improves Lead Times as well as makes quality problems easier and faster to be seen and corrected.

5. C.I. ---continuous improvement. This means not only monthly Kaizen events but Individual Daily Kaizens and an encouragement of operators to change things in the area they work so as to improve them even if that means producing some failures in the interim. It is from those failures that if managed correctly that better processes are ultimately found.

Recall the types of waste categories in Lean from the acronym:

TIIMWOOD

Transportation-movement of goods and products

Inventory- the collection of finished or partially completed products

Intellect- not fully using the reasoning power and intellect to your workers

Motion- movement of people

Waiting- any process step that is waiting on work to be delivered from the up-stream process

Overproduction- making more product that the customer wants. This is usually done to provide buffer due to inconsistency in the overall production process

Over processing- taking more steps than necessary to make a product

Defects- making bad items

Instill problem solving techniques in everyone at your facility by training and using PDCA/A3. The best way to get better at problem solving using this methodology is to continually do it. The more repetitions, just like in sports or most anything else, the better one becomes at it.

Leaders should be mentoring/coaching people in their charge on how to use A3's in following the PDCA cycle. Driving this down to everyone gets it closer and closer to the root cause of problems and that is where they are most easily solved as well as most inexpensively solved.

Institute daily or 2x daily Gemba Walks. Get rid of conference room meetings going over PowerPoint and Excel. Instead take the meetings to the floor in a Gemba Walk and see first-hand where the value added work is being done and provide support for the operators doing that value added work. If possible move all the leadership offices to gemba on the plant floor as well.

Instill the mindset that every day every single person from the plant manager on down to the operator and administrative assistant should be thinking of ways of improving the overall business. Train everyone in Lean principles so that they now have context in which to come up with ideas for improvement.

Schedule and carry out every month Kaizen events that are designed to improve a given area. Make sure everyone at the site cycles through Kaizen event. Act on and take notice to every Individual Daily Kaizen idea each and any operator may have on improving the business. Don't get frustrated if most of these ideas are old, been tried and failed or just plain bad. This is to be expected. But think of combining several aspects of bad ideas into a new good idea and that will make the program worthwhile.

If a plant that has never heard of Lean can start to do this every day, in 6 months they will look and feel a lot different. The workers will feel more involved and empowered. They will be more intellectually engaged and connected to one another by a holistic process. The instinctively will start thinking in Lean terms out of reflex.

Within 1 year the plant will no longer recognize itself and will wonder how they ever did business or functioned the old way.

Eventually if they even get good enough at this they can actually do consulting for other companies by allowing them to come to their site and see how Lean is implemented and actually do Lean training for paid services as a money making venture. I have seen this happen several times.

Lean is a journey. It is never completed. But it does have to start at some point. The good news is it takes little in the way of monetary investment in most cases.

Hopefully this book can convince a few naysayers and skeptics out there to take the steps to go down this journey.

GLOSSARY OF LEAN TERMS

WORD OR PHRASE	MEANING
5S	A way to detect "normal" from "abnormal" right away. The 5S's are: (1.) Sort (2.) Set (3.) Shine (4.) Standardize (5.)Sustain
A3	A size of paper that is meant to track and organize PDCA (Plan Do Check Act) problem solving such that the status of a problem can be readily seen and understood by anyone
Andon signal	A signal usually a light, in a Pull system that is meant to convey to an upstream process step that the next step is ready for product or supplies
Cell	The way to organize a logical work area usually in a U-shape as opposed to the traditional long manufacturing straight line in Western factories
Cycle time	The actual time it takes for a given process steop to run. Not to be confused with takt time or lead time
First pass yield	The amount of final product that makes it through the entire series of process steps with no defect. Eg--if 100 supply units enter a system and 90 good parts come out then the FTY= 90%
flow	Uninterrupted value added work. Information , materials and product moving through a system unimpeded.
Gemba	Where the work is done. As in the phrase "go to gemba" meaning go see for yourself where the work is done
Genchi Genbutsu	"go and see". A Toyota principle that managers, if they want to help solve problems on the shop floor need to go and help find solutions on the shop floor.
Hansei	Japanese word meaning "critical self reflection". In Lean it is an attitude that workers and managers should take where they readily admit and acknowledge their mistake and make a commitment to improve. Even when a project or problem is ultimately solve correctly there is a self reflection meeting to discuss things that did go wrong and how to improve. An important feature of the 5th Principle of Lean- C.I. Continuous improvement
Heijunka	Production leveling. A way to smooth out variation in production due to changing customer demand. E.g.--make a RED product on Monday from 2-6 each week and a BLUE product on Tue from 7-10, etc rather than making all RED till an order is complete then changing over to all BLUE till an order is complete etc. To make this work the principles of SMED (see below) must be mastered.
Hoshin Kanri	The overall strategic vision and objectives of a company and how that vision is used by the lower divisions and groups to derive their objectives in order to meet this vision
Inventory wait time	WIP or work in process parts or semi-parts that must wait in a que between process steps. This causes a back up and clogs the flow of a system
Jidoka	refers to automating a system so that the machine interacts smoothly with the human operators running a line
Kaikaku	Big change. This is in opposition to Kaizen which usually involves small incremental change done continuously. A project that products very large change is Kaikaku. This is usually the result of many many Kaizen changes
Kaizen	literally means "change for the good".
Kanban	A kanban means "card". In a Lean context it is a signal (historically it really was a card) from a downstream step to an upstream step letting them know that a part or material is needed. This aids in making a Pull system operatre effectively. Today rather than cards the signal is usually electronic
Lead time	The time between a customer order and fullfilling that order by deliver to the customer.
Milk Run	A dedicated predictable and consistent route a carrier or truck takes to pick up products and materials on a regular basis.
Monument	A large piece of equipment or object in a plant that is too large and expensive to move so a process is planned around it.
Muda	A type of waste. Non valude added work. Or anything that requires more resources than is needed.
Mura	a type of waste defined by unevenness or inconsitency in production.
Muri	waste from unreasonable demand. Usually reduced by understanding takt time and applying standard work
NEMAWASHI	Literally means "going around the roots". More loosely in Western parlance "laying the ground work". What is meant is that team leaders goes around and explains and gets consensus for a project on every leader or group that it touches. Comes from the Japanese gardening concept of digging around the roots of a tree to get it ready to remove and replace at another location.
OEE	Operational Equipment Efficiency. The mathematical product of the multiplication of a machine's "quality" , "availability", and "rate" Expressed in a percentage where the max is 100%. So a machine that makes 100% good products, and is never down (100% available) and running as fast as it can (100% rate) has an OEE of 100%. Can also be applied to a process as well as a piece of equipment.
OSKKK	An acronym mean to show Toyota's prioritization of improvement. O-observe a process. S-standardize the process as it stands. K-Kaizen for process improvement. K-kaizen for equipment improvement (new equipment). K-kaizen for plant improvement (new plant)
PDCA	A problem solving methodology of Plan, Do, Check Act, popularized by Deming
Poka Yoke	To "idiot proof" a solution. Make a process or part such that it cannot be put together or used wrong.
Pull	A way that a product, information etc is moved through a process that is diametrically opposed to "push". This helps control WIP (Work In Process)
SIPOC	Supplier Input Process Output Customer
SMED	Single Minute Exchange of Dye. A method to reduce change over time when a process goes from making one product to another.
Spaghetti map	A map that shows the actual foot traffic of operators or material as it moves from step to step of a process. Called a spaghetti map due to it's appearance like a plate of spaghetti due to all the non-linear travel geography
Standard Work	The way to reduce variation in a process and to establish a baseline upon which later improvements will be able to tell their impact. A very imortant problem solving tool in Lean
Takt time	The Pace of sales. How fast the csutomer buys the products of a process.
Toyota kata	Method of mentoring used at Toyota where a manager/supervisor guides someone t hrough the PDCA problem solving on a problem usually using an A3
Value add	Any activity that changes the form or function of a product or service that the customer is willing to pay for.
VSM	A detailed pictorial hand drawn map showing the process steps that deliver value to the customer
Water Spider	A person who works in a process put not assigned to a particular location or cell and is able to move around to a variety of cells in order to aid the flow and work overall.
WIP	Work In Process. Inventory between process steps.

OSKKK	An acronym mean to show Toyota's prioritization of improvement. O-observe a process. S-standardize the process as it stands. K-Kaizen for process improvement. K-kaizen for equipment improvement (new equipment). K-kaizen for plant improvement (new plant)
PDCA	A problem solving methodology of Plan, Do, Check Act, popularized by Deming
Poka Yoke	To "idiot proof" a solution. Make a process or part such that it cannot be put together or used wrong.
Pull	A way that a product, information etc is moved through a process that is diametrically opposed to "push". This helps control WIP (Work In Process)
SIPOC	Supplier Input Process Output Customer
SMED	Single Minute Exchange of Dye. A method to reduce change over time when a process goes from making one product to another.
Spaghetti map	A map that shows the actual foot traffic of operators or material as it moves from step to step of a process. Called a spaghetti map due to it's appearance like a plate of spaghetti due to all the non-linear travel geography
Standard Work	The way to reduce variation in a process and to establish a baseline upon which later improvements will be eable to tell their impact. A very imortant problem solving tool in Lean
Takt time	The Pace of sales. How fast the csutomer buys the products of a process.
Toyota kata	Method of mentoring used at Toyota where a manager/supervisor guides someone t hrough the PDCA problem solving on a problem usually using an A3
Value add	Any activity that changes the form or function of a product or service that the customer is willing to pay for.
VSM	A detailed pictorial hand drawn map showing the process steps that deliver value to the customer
Water Spider	A person who works in a process put not assigned to a particular location or cell and is able to move around to a variety of cells in order to aid the flow and work overall.
WIP	Work In Process. Inventory between process steps.

www.ingramcontent.com/pod-product-compliance
Lightning Source LLC
Chambersburg PA
CBHW060416190526
45169CB00002B/931